AF374819

God and Cosmos Theory

HyperCosmos
Second Kind HyperCosmos
Tetragrammaton

Stephen Blaha Ph. D.
Blaha Research

Pingree-Hill Publishing
MMXXIII

To Margaret

Some Other Books by Stephen Blaha

All the Megaverse! Starships Exploring the Endless Universes of the Cosmos using the Baryonic Force (Blaha Research, Auburn, NH, 2014)

SuperCivilizations: Civilizations as Superorganisms (McMann-Fisher Publishing, Auburn, NH, 2010)

All the Universe! Faster Than Light Tachyon Quark Starships & Particle Accelerators with the LHC as a Prototype Starship Drive Scientific Edition (Pingree-Hill Publishing, Auburn, NH, 2011).

Unification of God Theory and Unified SuperStandard Model THIRD EDITION (Pingree Hill Publishing, Auburn, NH, 2018).

The Exact QED Calculation of the Fine Structure Constant Implies ALL 4D Universes have the Same Physics/Life Prospects (Pingree Hill Publishing, Auburn, NH, 2019).

Integration of General Relativity and Quantum Theory: Octonion Cosmology, GiFT, Creation/Annihilation Spaces CASe, Reduction of Spaces to a Few Fermions and Symmetries in Fundamental Frames (Pingree Hill Publishing, Auburn, NH, 2021).

Passing Through Nature to Eternity ProtoCosmos, HyperCosmos, Unified SuperStandard Theory (Pingree Hill Publishing, Auburn, NH, 2022).

HyperCosmos Fractionation and Fundamental Reference Frame Based Unification: Particle Inner Space Basis of Parton and Dual Resonance Models (Pingree Hill Publishing, Auburn, NH, 2022).

A New UniDimension ProtoCosmos and SuperString F-Theory Relation to the HyperCosmos (Pingree Hill Publishing, Auburn, NH, 2022).

The Cosmic Panorama: ProtoCosmos, HyperCosmos, Unified SuperStandard Theory (UST) Derivation (Pingree Hill Publishing, Auburn, NH, 2022).

Ultimate Origin: ProtoCosmos and HyperCosmos (Pingree Hill Publishing, Auburn, NH, 2022).

UltraUnification and the Generation of the Cosmos (Pingree Hill Publishing, Auburn, NH, 2023).

Available on Amazon.com, bn.com, Amazon.co.uk and other international web sites as well as at better bookstores.

CONTENTS

Introduction

The book begins with a qualitative description of the features of the 42 Cosmos Theory spaces: the HyperCosmos, the Second Kind HyperCosmos, their 20 HyperUnification spaces, the Full HyperUnification space, and the UltraUnification space. The rationale for the origin and sequence of spaces is detailed.

The form of the 42 space-time dimension Full HyperUnification space is described in detail including its four levels (Tetragrammaton) and its three parts. (These three parts are united in the UltraUnification space.)

The book explores experimental data supporting the existence of Cosmos Theory. The generation of Cosmos Theory from one initial dimension is also considered.

Then the book describes five qualitative characteristics of Cosmos Theory. The twelve Physical characteristics of God are described next. The Physical characteristics of Cosmos Theory are compared to the Physical attributes of God. The book concludes Cosmos Theory may be viewed as an analogue to God's physical features: timeless, omniscience, Also Cosmos Theory contains numbers associated with God: 3, 4, 32, 42 and 88. These numbers are shown to be inherent in Cosmos Theory. They are consequences of the choice of ten HyperCosmos spaces where each space is defined with Cayley-Dickson hypercomplex numbers. Ten is also a number associated with God.

The 42 space-time dimension space exhibits a 1, 4, 9 sequence of blocks of dimensions, which are the squares of 1, 2, 3. These numeric relations have appeared in various contexts.

The depth of the theory and its embodiment of God-like features suggest it should be considered as a candidate for the much talked about Theory of Everything. Cosmos Theory is architecture (from the Divine Architect?) for the Cosmos; God creates universes according to this theory (which He alone must have originated.) It is not a surprise that Cosmos Theory manifests some of God's attributes. Mankind is often viewed as having been formed in God's image.

The theory is formulated so all features can be generated from one primordial dimension in the UltraUnification space.

1. Cosmos Theory

Intimations of God appear in the author's Cosmos Theory.[1] This chapter describes aspects of Cosmos Theory. It consists of 42 spaces with well-defined structures for fundamental particles and their interactions. This chapter is a prelude to the discussion of its possible relation to God.

1.1 Origin of Cosmos Theory

Cosmos Theory originated as a major extension of the Standard Model of Elementary Particles. It began some years ago as a generalization of the Standard Model which the author called the Unified SuperStandard Theory (UST). In the autumn of 2019 the author noticed that a theory based on hypercomplex (Cayley-Dickson) numbers: quaternions, octonions, had the same contents as the UST. From that point the theory evolved into HyperCosmos Theory in 2020 – 2023.[2] HyperCosmos Theory has a set of ten spaces. One of the spaces is for our universe and describes the UST.

Recently the author generalized the framework of the theory to Cosmos Theory, which contains 42 spaces: ten HyperCosmos spaces, 10 spaces of a related HyperCosmos of the Second Kind, 20 HyperUnification spaces, one 42 space-time dimension Full HyperUnification space and one UltraUnification space. See Fig. 1.1 for a diagram of the 42 highly interrelated spaces of the Cosmos. Our universe has an N = 7, level 4 HyperCosmos space.[3]

Each space has a set of dimensions in an array called the dimension array for space-time and for internal symmetry groups. The set of dimensions are organized into a *square* array of dimensions that we call a *dimension array*. Dimensions are described in detail in Blaha (2023a). We will think of dimensions as similar to the four space-time dimensions of our universe. Internal symmetry groups use dimensions in a somewhat similar way in their irreducible representation definitions.

The dimension array of each space specifies the number of fundamental fermions (half integer spin particles); the number of gauge vector fields, the number of scalar bosons including Higgs bosons; the space-time dimensions; and the set of internal symmetry groups (interactions).

1.2 Guiding Principles of Cosmos Structure

The 42 spaces of Cosmos theory are independent but they are tightly intertwined. The creation of Cosmos theory is based on the fermions—half-integer spin fundamental particles (quarks and leptons) of which all matter is composed in the universes of all spaces. Fermions are defined in terms of creation and annihilation operators that appear in their definitions. These operators create and annihilate particles.

---

[1] See Chapter 9 of Blaha (2023a) "*UltraUnification and the Generation of the Cosmos*" listed in the References section.

[2] See the 2021-2022 Blaha books listed in the References section.

[3] If Dark matter does not exist then our universe has the features of a Second Kind HyperCosmos space.

We will call these operators the b's and d's respectively in the way that they are symbolized in their definitions for each space.

We will now describe the principles that guided the development of the form of Cosmos Theory. Then we briefly describe the origin and role of each space.

1.2.1 The b's and d's of a Cosmos Space

In each space a fermion's b's and d's may be defined. One starts with a definition particular to the space with a certain number of spins. (See Fig. 1.2) Then we generalize the b's and d's to be hypercomplex numbers. Each space's number of space-time dimensions determines the Cayley hypercomplex number n shown in Fig. 1.2.

The set of b's and d's is thereby determined for each space. The number of b's and d's – space by space – is shown in the d_{dN} column of Fig. 1.2. These numbers also appear in the energy spectrum of the ProtoCosmos Model. Thus the general form of each space's dimensions is determined. Then each set of dimensions is subdivided into subsets for the internal symmetry groups and also into subsets of fundamental fermions (quarks and leptons such as electrons and neutrinos). Each space is thus "fully" defined.

The existence of Dark Matter is an open question at present. The HyperCosmos spaces treat Dark Matter on an equal footing with "Normal" Matter. We introduce a new set of 10 spaces called HyperCosmos spaces of the Second Kind that is the "same" except that they do not support Dark Matter with consequent changes in some numerical parameters. The Second Kind HyperCosmos naturally appears in the 42 space-time dimension space.

1.2.2 Fundamental Reference Frames (FRFs)

The guiding principle here is to reduce each set of b's and d's of each space to a minimal set. This is accomplished by requiring the b's and d's of a "normal" General Relativistic reference frame (Its x, y, z, …, and t) to be the result of a General Relativistic – Internal Symmetry transformation from its FRF. This process is analogous to the Special Relativistic transformation from a particle's momentum "rest frame" where it only has one non-zero coordinate (energy) to a moving frame where it may have all four momentum coordinates with non-zero values.

The FRF of a space has a number of non-zero dimensions in it. It is equal to the number of b's and d's in the space's fermion before that number is expanded to hypercomplex b's and d's in section 1.2.1 above.

It is possible to define a transformation that maps one FRF non-zero dimension to all the dimensions in the dimension array for a space.

1.2.3 The HyperUnification Space for each HyperCosmos Space

The transformations in section 1.2.2 combine General Relativistic and Internal Symmetry. We now define a HyperUnification space for each HyperCosmos space and Second Kind HyperCosmos space that has *purely* General Relativistic transformations that transform ("rotate") dimension arrays. A HyperUnification space supports these

transformations because it is a well-defined[4] higher space-time dimension space. The r'
column of Fig. 1.2 shows the higher space-time dimension HyperUnification space for
each HyperCosmos space.[5] The b's and d's of each space *determine* the space-time
dimension of its HyperUnification space.

A HyperUnification space gives true unification of space-time and Internal
Symmetries for the corresponding HyperCosmos space.

1.2.4 The Full HyperUnification Space for the set of 20 HyperCosmos and Second Kind HyperCosmos Spaces

It is possible to combine the twenty HyperUnification spaces of the
HyperCosmos and the HyperCosmos of the Second Kind in a single 42 space-time
dimension space of the form of a HyperCosmos space. The set of HyperUnification
spaces is strung along a diagonal in the dimension array of the 42 space-time dimension
space. The dimension array of this space contains the set of HyperUnification spaces'
dimension arrays.[6]

Some of the transformations of the dimensions of this 42 space-time dimension
space may be viewed as a combination of the transformations of each of the 20
HyperUnification spaces. The 42 dimension space transformations are not pure General
Relativistic transformations. They are a not-unified subset of the General Relativistic
transformations of the spaces.

1.2.5 88-Space-time Dimension UltraUnification Space

A unification of the contents of the 42 space-time dimension space can be made
in a similar fashion to the HyperUnification of HyperCosmos and Second Kind
HyperCosmos as specified by Footnote 4 above. The combined 20 HyperUnification
spaces are displayed in a square array within a dimension array of the UltraUnification
space as shown in Fig. 1.3. The UltraUnification space has a space-time defined by
requiring its transformations to be purely General Relativistic transformations in the
UltraUnification space. As a result the various HyperUnification blocks within the 42
dimension space are mixed together—thus unified as one composite whole.

Again, the unification framework is based on transformations of the fermion b's
and d's.

---

[4] The choice of the space-time dimensions of a HyperUnification space is based on the need to construct a space
whose space-time accommodates a pure General Relativistic transformation that mixes all dimensions of the target
space dimension array expressed as a vector. Thus the number of components in the target space dimension array is
the length of the HyperUnification vector being transformed. This length specifies the number of space-time
dimensions of the HyperUnification space. See Fig. 1.2 where the number of space-time dimensions of the
HyperUnification space r' is listed for each HyperCosmos space of space-time dimension r. This specification is
motivated by the equivalent transformations of fermion b's and d's.

[5] Equation 4.8 of Blaha (2023a). Its justification is The defining feature of a HyperUnification space with space-time
dimension r' of a space with space-time dimension r is that the number of components in the dimension r' GR
transformation vector equals the number of components d_{dN} in the dimension array of the space of space-time
dimension r.

[6] See Blaha (2023a). The sum of column lengths of the 20 HyperUnification vectors plus 8 equals the column length
of the 42 space-time dimension dimension array. See footnote 4.

1.2.6 Beyond UltraUnification

Being totally unified in UltraUnification space there is no reason to go beyond it bringing Cosmos Theory to closure.

1.3 Structure of the Cosmos Theory

The structure of the theory is somewhat complex. We view the theory as a paradigm or "blueprint" for the Cosmos. One or more universes may exist for each of the 42 spaces. Universes may be interrelated by interactions (forces). Consequently Multiverses are possible.

1.4 HyperCosmos Spaces

We now turn to consider each type of space.

There are ten HyperCosmos spaces listed in Fig. 1.2. A universe of a given space type has a certain number of space-time dimensions. (Our universe has 4.) It also has a certain number of dimensions for internal symmetries (interactions). Our universe has 252 such dimensions for SU(3), SU(2)⊗U(1), … group interactions.

If our universe has a Dark Matter sector similar to the Normal Matter sector then it has the structure of an N = 7 HyperCosmos space as shown in Fig. 1.2.

1.5 Second Kind HyperCosmos Spaces

If our universe has no Dark Matter sector then it has the structure of an N = 7 Second Kind HyperCosmos space. Second Kind HyperCosmos spaces have no Dark Matter sector. Their dimension arrays are exactly ½ of the corresponding HyperCosmos space dimension array. See Blaha (2023a) for details.

1.6 HyperUnification Spaces

There are 20 HyperUnification spaces: ten for the HyperCosmos spaces and ten for the Second Kind HyperCosmos spaces. They are constructed according to the rules given in section 1.2.3 above. The transformations a HyperUnification space can take an FRF with only one non-zero dimension and generate the dimension array of its corresponding HyperCosmos space. See Blaha (2023a) for details.

1.7 Full HyperUnification Space – 42 Space

The Full HyperUnification space is defined to hold the 10 HyperCosmos HyperUnification spaces plus the 10 Second Kind HyperCosmos HyperUnification spaces. The minimal space[7] that will hold the 10 HyperCosmos HyperUnification spaces has 42 space-time dimensions. (We require the Full HyperUnification space to have the same form as the HyperCosmos spaces but with a higher space-time dimension.)

The 42 space-time dimension Full HyperUnification space dimension array holds both the 10 HyperCosmos HyperUnification spaces and the 10 Second Kind HyperCosmos HyperUnification spaces dimension arrays (and the 42 space-time dimension General Relativistic transformations).

---

[7] A space with smaller than 42 space-time dimensions will not hold the dimension arrays of the 10 HyperCosmos HyperUnification spaces.

The Full HyperUnification space total dimensions expressed as a square array has 8,388,608 by 8,388,608 components. See Fig. 1.3. The 10 HyperCosmos HyperUnification spaces' dimension arrays occupy the 5,592,400 by 5,592,400 square array components block labeled by "A". The 10 Second Kind HyperCosmos HyperUnification spaces' dimension arrays occupy the 2,796,200 by 2,796,200 square array components block labeled by "B". The remaining 8 by 8 square array block labeled by "C" fill out the Full HyperUnification space dimension array.

For both A and B blocks, the HyperUnification space dimension blocks within them are strung along the diagonal as shown in Fig. 1.3. See Fig. 1.1 for the overall Cosmos Theory design.

Note that the length of block B, 2,796,200, is one half of the length of block A, 5,592,400. The difference is due to block A having both a 2,796,200 by 2,796,200 subblock for its Normal Matter sector and a 2,796,200 by 2,796,200 subblock for its Dark Matter sector. Block B only has a Normal Matter sector. Thus we see that the 42 space-time dimension array has three parts. This fact will be of interest later.

The 42 space exhibits a 1, 4, 9 sequence of blocks of dimensions. They are the squares of the 1, 2, 3 sequence of the A and B block sides and the total 42 space dimension array block side 2,796,200, 5,592,400, 8,388,608. These numeric relations have appeared in various contexts.[8]

1.8 UltraUnification Space – 88-Space

The Full HyperUnification space has a set of combined General Relativistic and block symmetry transformations that transform the array of Fig. 1.3 into a different form in a different 42 space-time dimension reference frame.

We achieve a further unification to *strictly* General Relativistic transformations in an 88-Space-time dimension space that we call the *UltraUnification space* or *"88-Space"*. The procedure is the same followed in section 1.2.3 and footnote 5 above.

The transformations of 88-Space can take an FRF with only one non-zero dimension and generate the dimension arrays of all the other spaces of Cosmos Theory. See Blaha (2023a) for details.

See Appendix 1-A for possible experimental support for the HyperCosmos. Appendix 1-B describes processes associated with Cosmos Theory.

1.9 Additional Spaces?

Having fully unified the Cosmos Theory we see no further need for spaces. Fig. 1.1 displays the four levels of Cosmos Theory.

1.10 Dirac and Eddington Large Number Conjectures

The 42 and 88-Space-time dimension of spaces evokes the conjectures of Dirac and Eddington who noted that physical constants and parameters can be combined to form a number of dimensionless constants with approximate values of 40 and 80. This seeming coincidence raises the question of a deep connection of the 42-Space and 88-Space and the 40 and 80 numeric values.

---

[8] Examples are the size of the monoliths in the movies *2001* and *2010*.

Level

1   88 Dimension
UltraUnification
Space

42 Dimension
Full HyperUnification
Space

2

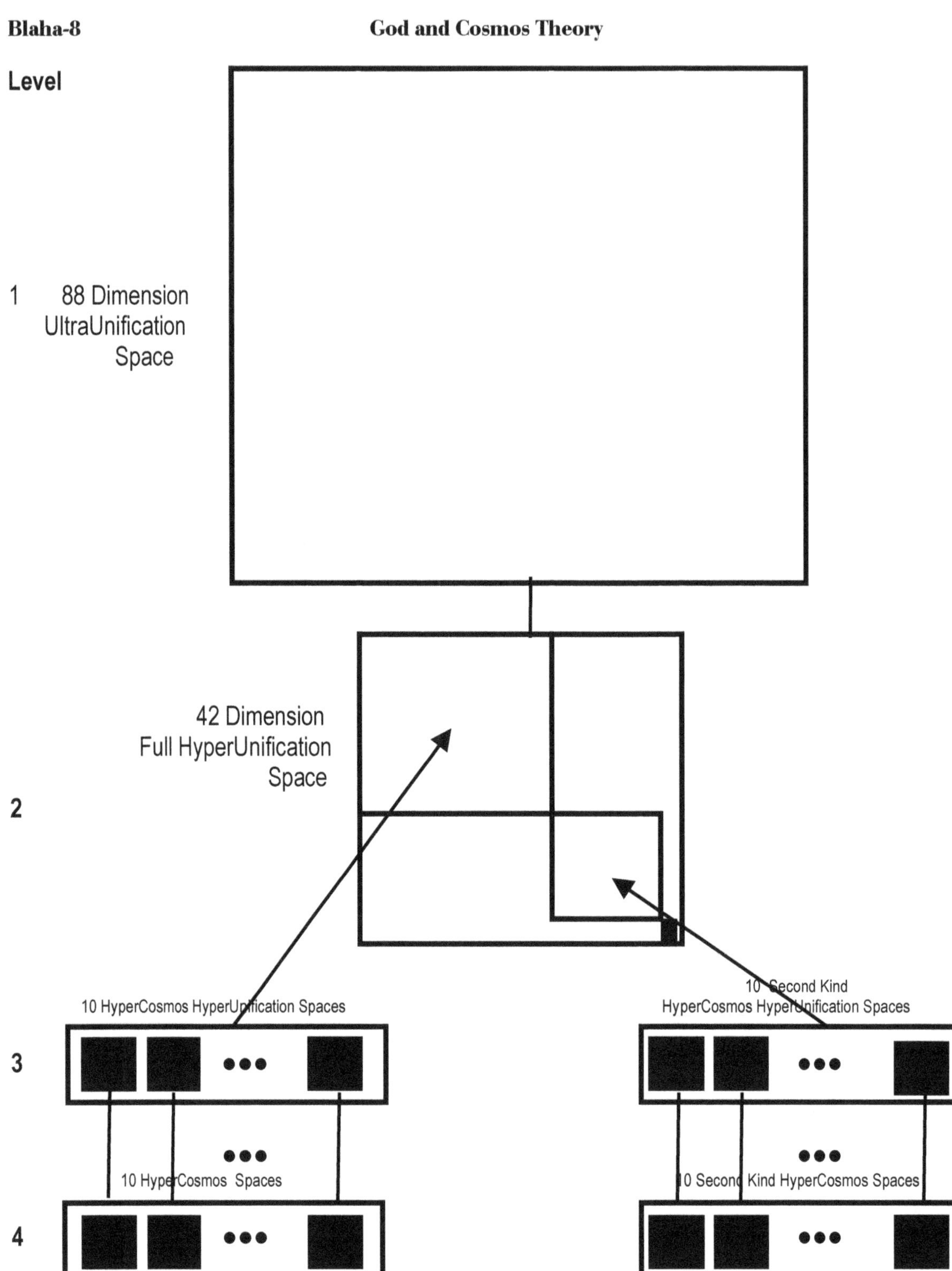

Figure 1.1. Diagram of the four levels of the 42 Cosmos Theory spaces. Not drawn to scale.

THE HYPERCOSMOS SPACES SPECTRUM

Blaha Space Number $N = O_s$	Cayley-Dickson Number n	Cayley Number d_c	Dimension Array column length d_{cd}	Dimension Array Size d_{dN}	Space-time-Dimension r	Higher Space-time r^t Source of d_{dN} r^t	Higher Space Array $d_{dN}{}^t$
0	10	1024	2048	2048^2	18	40	2^{44}
1	9	512	1024	1024^2	16	36	2^{40}
2	8	256	512	512^2	14	32	2^{36}
3	7	128	256	256^2	12	28	2^{32}
4	6	64	128	128^2	10	24	2^{28}
5	5	32	64	64^2	8	20	2^{24}
6	4	16	32	32^2	6	16	2^{20}
7	**3**	**8**	**16**	**16^2**	**4**	**12**	**2^{16}**
8	2	4	8	8^2	2	8	2^{12}
9	1	2	4	4^2	0	4	2^8

Figure 1.2. The HyperCosmos spaces spectrum related to their HyperUnification space's space-time dimension and dimension array.

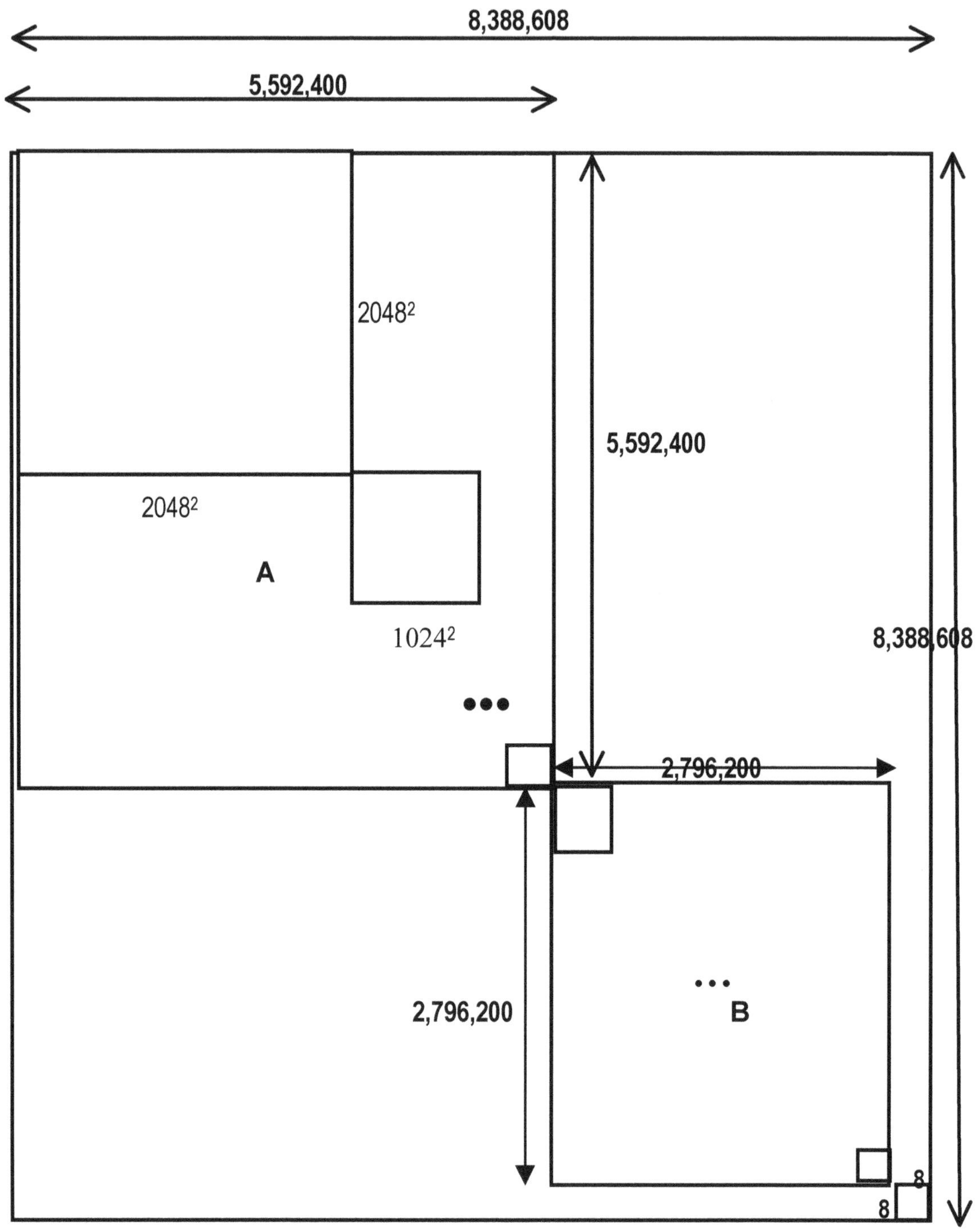

Figure 1.3. Form of a 42 space-time dimnesion Full HyperUnification transformation. It is also the form of the dimension array of 42 dimension space-time. All blocks are square. The figure is not drawn to scale.

Appendix 1-A. Experimental Support for the HyperCosmos and UST

There is evidence for the ProtoCosmos and the HyperCosmos. It is not direct evidence since we are bound by our universe and its elapsed lifetime. It is evidence that indirectly supports their existence and their structure.

1. Direct Generation of the HyperCosmos spectrum from a UniDimension ProtoCosmos.

2. Relic Structure in Internal Symmetries and the Fundamental Fermion Spectrum. The Standard Model internal symmetry $SU(2) \otimes U(1) \otimes SU(3)$ is shown to appear in HyperCosmos/QUeST as a natural separation not requiring symmetry breaking Higgs fields. QUeST introduces additional symmetries that add to those of the Standard Model in a direct way—again without Higgs bosons. Higgs breaking is restricted to ElectroWeak-like sectors.

3. *The separation of fermions and symmetries into superluminal and subluminal parts in the Fundamental Reference Frame leads to the fundamental fermion spectrum, to the set of internal symmetries, and to the understanding of the relation of the Higgs Mechanism and Strong Interaction Confinement (the Higgs-Confinement Dichotomy).*

4. The Dual Resonance Model of hadron scattering is *directly* based on the HyperCosmos using fractionated hadrons. This approach is an alternative to String Theories. This approach supports hadronic Veneziano amplitudes.

5. The Parton Model of Deep Inelastic Lepton-Nucleon Scattering may be viewed as based on fractionated particles. One finds that it "explains" quasi-free partons, the apparent lack of strong interactions between partons thus giving parton freedom, and scaling in x, as well as other features.

6. The HyperCosmos spaces spectrum can be cast into a form similar to the Regge trajectories of scattering amplitudes.

7. The fractionated particle formalism has lattice theory as a restricted special case.

8. The HyperCosmos leads to a deep unification based on its Fundamental Reference Frame formalism. For example, in the case of our universe, there is a unified 12 space-time dimension formalism that generates the unified General Relativity and Internal Symmetries of our universe.

9. The space-time dimensions, 42 and 88, as well as the numbers 3, 4, 10, and 32 that appear within Cosmos Theory appear to indicate that Cosmos Theory may be at the deepest level and may reflect attributes of God. See Chapter 4.

10. The 42-Space exhibits a 1, 4, 9 sequence of blocks of dimensions. They are the squares of the 1, 2, 3 sequence of the A and B block sides and the total 42 space dimension array block side 2,796,200, 5,592,400, 8,388,608. These numeric relations have appeared in various contexts.

Appendix 1-B. The Generation of the Cosmos

Ideally the Cosmos may be viewed as created from nothing and then evolving according to the dictates of "inexorable" Physical Law. Our Cosmos almost meets that goal. The Prime Mover (God) starts from one dimension—a primitive term, undefined except that it has no value or substance—treatable as a "mental" concept. The dimension exists in the Fundamental Reference Frame (FRF) of a space with 88 space-time dimensions whose dimensions simply specify coordinates.

Treating this space as the HyperUnification of a 42 space-time dimension space we define a set of 88 space-time dimension General Relativistic (GR) transformations that take a vector of d_{cN_H} components (the number of elements in the 42 dimension array of dimensions; See eq. 6.8a.) with only one non-zero component (the original dimension) and generate a vector with d_{cN_H} non-zero components (dimensions).

One might ask if the transformation creates dimensions. In a physical sense the answer is no. The GR transformation "merely" relates the "situation" in the FRF to the "situation" in the destination reference frame. In the 1970s it became clear that the particles seen in one reference frame are equivalent to a different set of particles in an accelerating reference frame. No particles were created. Reference frames determine what one sees in general.

The above process does have a result: the conceptual expansion of the one original dimension to the dimensions of the second, third and fourth levels of Fig. 9.1. And so the Prime Mover has generates, without creating, the framework of the Cosmos.

Having populated the 42 dimension Full HyperUnification space dimension array, (chapters 6 and 7) we see that each HyperCosmos space and each Second Kind HyperCosmos space has its HyperUnification space dimension array populated. The dimensions in each HyperUnification space dimension array may be used to define the dimension array of the corresponding HyperCosmos and/or Second kind HyperCosmos space.

Up to this point nothing has been created. All dimensions flow from the original one dimension in the 88 dimension UU space. The structure of the Cosmos has been *conceptually* generated from one original dimension. And that is all that is needed to support Creation.

1-B.1 Structure of the Cosmos

The structure of the UU space and the spaces on levels 2 – 4 determines the overall structure of the HyperCosmos and Second Kind HyperCosmos spaces. The FRF contents are specified in chapters 4 and 5 by considering the forms of symmetry groups and fermions implied by the two time coordinates generated through su(1,1): a group that appears in all spaces. After a General Relativistic transformation in each space's HyperUnification space the dimension array of each space emerges. The dimension array of each HyperCosmos and Second Kind HyperCosmos space consists of replicates of its FRF content. The pattern (splittings) of internal symmetries of each space is thereby determined. The fundamental fermions of each space are similarly determined as are vector gauge particles and scalar Higgs and non-Higgs bosons.

Thus the four levels of spaces in Fig. 9.1 fully determine the symmetry groups and particles associated with each space.

1-B.2 Creation of Universes – ProtoCosmos

Given the definition of spaces and their contents derived above we can turn to the creation of universes.[9] The UniDimension ProtoCosmos Model was shown in Blaha (2022g) to be derivable from one dimension where the dimension is fractionated[10] into an infinite set[11] of "dust." The dust which is formless may be shaped into a line segment [-1, 1] and transformed to a continuous line. A dynamics may be introduced consisting of a one-dimension Dirac equation with a dimensionless 1/x potential. The spectrum of this equation can be shaped by setting constants to produce a set of eigenfunctions: each with a fermion wave function factor. Thus each HyperCosmos space[12] can be instantiated by creating states. These states specify universes that begin with a certain energy and dimension array.

The ProtoCosmos generates the spaces and universes of the HyperCosmos from one dimension. By making the ProtoCosmos energy spectrum consistent with the HyperCosmos (and/or Second Kind HyperCosmos) spectrum we can *consistently* specify the one dimension of the UU FRF as the *source* of the ProtoCosmos one dimension in order to create a full Cosmos.

The result is a Cosmos populated with spaces and universes containing symmetries and matter. By universes we mean instances of spaces. This definition includes our universe, possible Multiverses, and universes containing universe.

The initial creation of universes seems to require the specification of universe states using the ProtoCosmos wave function formalism. Subsequently, if universes have interactions and a Quantum Field Theory formulation[13] they may generate additional universes through collisions. They may also generate subuniverses. The author's previous books describe these possibilities.

1-B.3 Evolution in Each Universe

When a universe is created in an enclosing universe it appears likely that it was created with a very large amount of internal energy at a point. Thermodynamics tells us that it will expand. This author developed a Big Bang theory describing the expansion process and the resultant Hubble expansion.[14]

---

[9] We take universes to mean an instantiation of a space. A universe may be our universe, a multiverse, or a set of interrelated universe within a host space.

[10] See Blaha (2022d).

[11] The reduction of a dimension to dust allows the Prime Mover to create the Cosmos from the "nearest thing" to Nothing.

[12] Or alternately for each Second Kind HyperCosmos space.

[13] See Blaha (2021f) and (2022c) for example.

[14] See Blaha (2004) and (2021d).

There is no need for a Prime Mover after universes (and the Cosmos of Fig. 9.1) appear. The universes evolve by the strict laws of dynamics generated from the symmetry group forms of the spaces and the particles specified from the FRFs of the HyperCosmos spaces.

1-B.4 Dynamical Laws of a Universe

The dynamical laws of a universe are specified by:

1. Dirac equations for fermions; and Yang-Mills equations for vector bosons with the details of the equations are determined by the relevant symmetry groups. General Relativistic generalizations of the equations are immediate. Scalar boson particles are also determined from the FRF of the universe's HyperCosmos space. See chapters 4 and 5. See chapters 12 – 14 of Blaha (2022f) for the PseudoQuantum formulation of the Riemann-Christoffel curvature tensor whose terms lead to the Euler-Lagrange formulation of the dynamical equations.

2. A PseudoQuantum formulation of the wave functions, states, and dynamical equations.

1-B.5 Conclusion

We conclude from the above discussion that all is based on dimension—the one primordial dimension resident in the FRF of the UltraUnification space.

2. The Nature of Cosmos Theory

Cosmos Theory specifies the form and features of the spaces of the Cosmos. It is a theoretical construct. There is no creation (or annihilation) within the theory.

One or more universes may be created for each Cosmos Theory space. The author's previous books (such as the 2021 book, *Universes are Particles*) specify the form of the wave functions used to define universes. Universes are created at points with an initial energy. They evolve by Hubble-like expansion. If universe interactions exist then Multiverses are possible. It is also possible for a universe to be created within a parent universe.

Some features of Cosmos theory and the universes that may be defined within its framework are:

1. There is no time or space outside universes. Each universe has its own time and space within it.

2. Matter and energy only exist within universes. The initial energy of universes created in the ProtoCosmos is specified in the ProtoCosmos space.

3. The evolution of a universe and its contents is solely determined by its dynamical equations which stem from the definition of interactions by the space's definition.

4. The contents of each universe (its types of particles and interactions) can be generated by a General Relativistic transformation from one dimension in the Fundamental Reference Frame (FRF) of its HyperUnification space.

5. The dimension arrays of all spaces can be generated by a General Relativistic transformation in 88-Space from one dimension in its Fundamental Reference Frame (FRF).

3. The Nature of God

Our knowledge of God may be viewed as composed of two parts: the humanistic part concerned with the relation of God and Mankind, and the "physical" part concerned with physical properties. This chapter considers physical aspects of God only:

General Physical Attributes:
1. God is not governed by time in the sense that He sees all time: past, present and future. Atemporal. No beginning or end.

2. God is not present at any spatial location—He is everywhere at all times.

3. God creates universes.

4. God sees all things as if He were viewing an infinite videotape of infinite detail.

5. God is both inside and outside of Cosmos Theory universes.

Quantitative Attributes:
1. Monotheistic God.

2. God is associated with a "four-foldness". Tetragrammaton.

3. God is viewed as having three parts by many.

4. The number 42 is associated with God.

5. The number 32 is often associated with God.

6. Some associate 88 with Godliness or "good Luck."

7. Ten is associated with God.

4. Cosmos Theory as an Analogue of God

We have suggested that Cosmos Theory is a possible analogue of God. Cosmos Theory is a paradigm or "blueprint" of the structure of the spaces of the Cosmos. It is not real. We see God as "creating" the blueprint as the Architect. Then God creates (instantiates) universes that evolve according to the Physical Laws embodied in the blueprint (with the possibility of Divine intervention not excluded.)

It is often thought that a paradigm reflects features of its creator. (Mankind is often viewed as created in God's image.) In that spirit we suggest that Cosmos Theory reflects features of God and call it an analogue of God.

We now do a point by point comparison of God's Physical features and Cosmos Theory features. The comparison clearly reflects analogous Physical features. Following the list of chapter 4 we find:

General Physical Attributes:
1. God is not governed by time in the sense that He sees all time: past, present and future. Atemporal. No beginning or end.

   Cosmos Similarity:
   No time exists outside of universes.

2. God is not present at any spatial location—He is everywhere at all times.

   Cosmos Similarity:
   No spatial distances exist outside of universes.

3. God creates universes.

   Cosmos Similarity:
   Same – Universes are created.

4. God sees all things as if He were viewing an infinite videotape of infinite detail.

   Cosmos Similarity:
   The beginning to end of a universe is specified dynamically.

Quantitative Attributes:
1. Monotheistic God.

   Cosmos Similarity:
   Cosmos.Theory unification.

2. God is associated with a "four-foldness". Tetragrammaton.

   Cosmos Similarity:
   Four levels of the Cosmos. Fig. 1.1.

3. God is often viewed as having a three-foldness also.

   Cosmos Similarity:
   The 42 dimension space has three equal parts. See Fig. 4.1. These parts are mixed together under 88-Space General Relativistic transformations. One may then view the three parts as unified to form one entity, a God analogue, based on 88-Space.

4. The number 42 is associated with God.

   Cosmos Similarity:
   42 spaces with a 42 dimension level 2 Full HyperUnification space

5. The number 32 is often associated with God.

   Cosmos Similarity:
   The ten HyperCosmos are associated with 32 other Cosmos spaces.

6. Some associate 88 with Godliness or "good Luck."

   Cosmos Similarity:
   Cosmos Theory has an 88-Space-time UltraUnification space.

The above cited numbers: 3, 4, 42, 32, and 88 all follow from the choice of 10 spaces as HyperCosmos spaces based on Cayley-Dickson numbers. The number 10 also has God-related significance. See Fig. 1.2.
In addition, the 42 space exhibits a 1, 4, 9 sequence of blocks of dimensions. They are the squares of the 1, 2, 3 sequence of the A and B block sides and the total 42 space dimension array block side: 2,796,200, 5,592,400, 8,388,608. These almost exact numeric relations have appeared in various contexts. See Fig. 1.3.

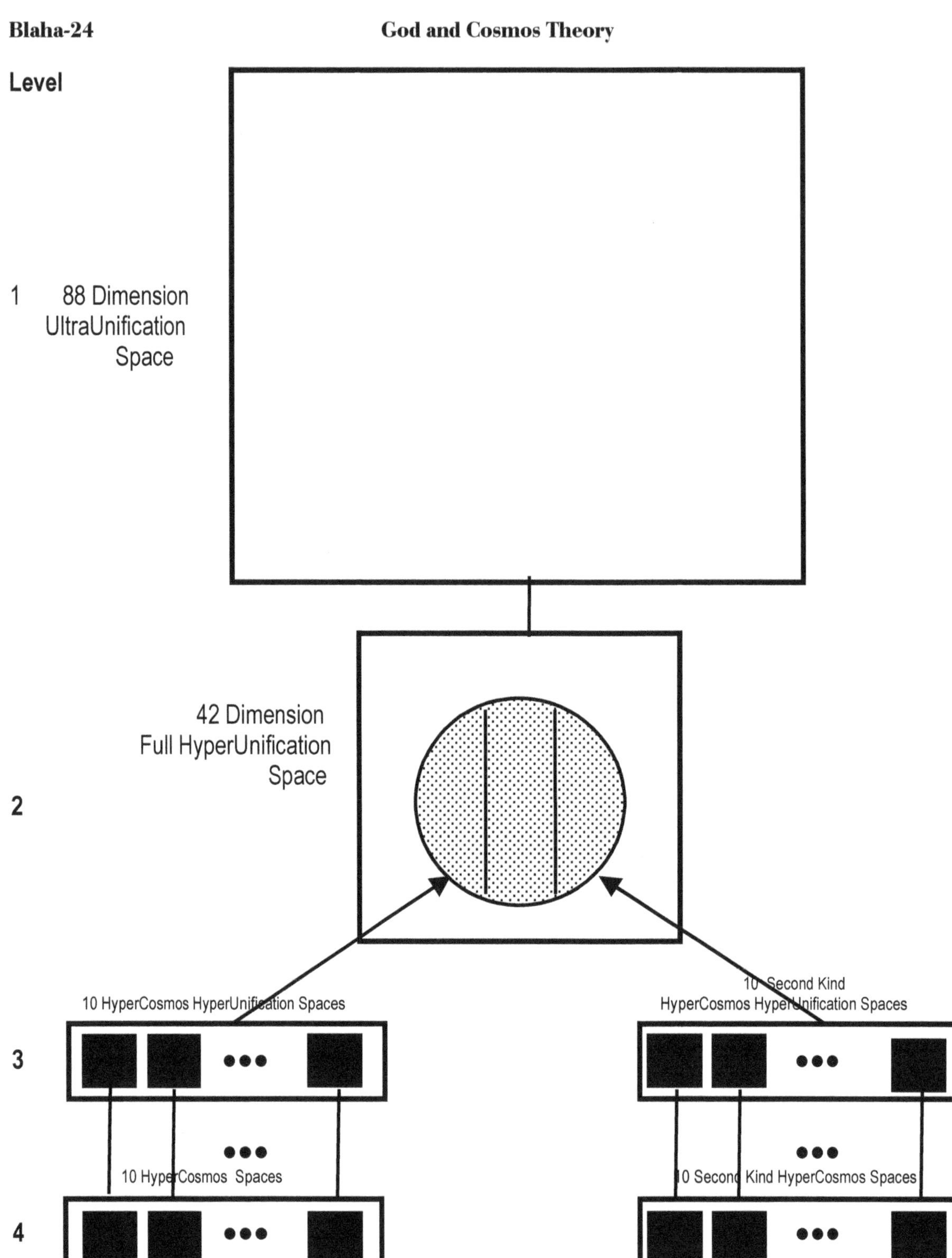

Figure 4.1. Diagram of the four levels of the Cosmos Theory spaces with level 2 displaying the three equal parts within blocks A and B of the Full HyperUnification space.

5. 88-Space as an Analogue of an Ultimate Brain

The dimension array of 88-Space contains an extremely large number of dimensions: $2^{92} = 4.95176*10^{27}$. Each dimension in 88-Space has an associated coordinate. The set of dimensions can be viewed as specifying points in a space. Thus they provide connections between points.

The synapses in a brain link neurons. A human brain has of the order of 10^{15} synapses. If we take dimensions to be the analogue of synapses then the total 88-Space dimensions is equivalent to the synapses of about $5*10^{12}$ or 5,000 billion human brains. An instrument of that size would be a formidable computer far beyond current or envisioned computers.

If Cosmos Theory is an analogue of God, then one gets a sense of the enormous power of the Godhead. God is without limit in the view of most people. "Practically" one can view 88-Space's number of dimensions as infinite.

We conclude Cosmos Theory has sufficient features to enable it to be thought an analogue of the Physical features of God.

SOME REFERENCES

Blaha, 1998, *Cosmos and Consciousness* (Pingree-Hill Publishing, Auburn, NH, 1998 and 2002).

_______, 2004, *Quantum Big Bang Cosmology: Complex Space-time General Relativity, Quantum Coordinates*,™ *Dodecahedral Universe, Inflation, and New Spin 0, ½, 1 & 2 Tachyons & Imagyons* (Pingree-Hill Publishing, Auburn, NH, 2004).

_______, 2010a, *Operator Metaphysics: A New Metaphysics Based on a New Operator Logic and a New Quantum Operator Logic that Lead to a Mathematical Basis for Plato's Theory of Ideas and Reality* (Pingree-Hill Publishing, Auburn, NH, 2010).

_______, 2010c, *SuperCivilizations: Civilizations as Superorganisms* (McMann-Fisher Publishing, Auburn, NH, 2010).

_______, 2018c, *God Theory* (Pingree Hill Publishing, Auburn, NH, 2018).

_______, 2018d, *Immortal Eye: God Theory: Second Edition* (Pingree Hill Publishing, Auburn, NH, 2018).

_______, 2018e, *Unification of God Theory and Unified SuperStandard Model THIRD EDITION* (Pingree Hill Publishing, Auburn, NH, 2018).

_______, 2020f, *A Very Conscious Universe* (Pingree Hill Publishing, Auburn, NH, 2020).

_______, 2021d, *Universes are Particles* (Pingree Hill Publishing, Auburn, NH, 2021).

_______, 2021e, *Octonion-like dna-based life, Universe expansion is decay, Emerging New Physics* (Pingree Hill Publishing, Auburn, NH, 2021).

_______, 2021g, *Quantum Space Theory With Application to Octonion Cosmology & Possibly To Fermionic Condensed Matter* (Pingree Hill Publishing, Auburn, NH, 2021).

_______, 2021h, *21ˢᵗ Century Natural Philosophy of Octonion Cosmology , and Predestination, Fate, and Free Will* (Pingree Hill Publishing, Auburn, NH, 2021).

_______, 2021i, *Beyond Octonion Cosmology II : Origin of the Quantum; A New Generalized Field Theory (GiFT); A Proof of the Spectrum of Universes; Atoms in Higher Universes* (Pingree Hill Publishing, Auburn, NH, 2021).

______, 2021j, *Integration of General Relativity and Quantum Theory: Octonion Cosmology, GiFT, Creation/Annihilation Spaces CASe, Reduction of Spaces to a Few Fermions and Symmetries in Fundamental Frames* (Pingree Hill Publishing, Auburn, NH, 2021).

______, 2022a, *New View of Octonion Cosmology Based on the Unification of General Relativit and Quantum Theory* (Pingree Hill Publishing, Auburn, NH, 2022).

______, 2022b, *The Gold Dust Beneath Hypercomplex Cosmology* (Pingree Hill Publishing, Auburn, NH, 2022).

______, 2022c, *Passing Through Nature to Eternity: ProtoCosmos, HyperCosmos, Unified SuperStandard Theory* (Pingree Hill Publishing, Auburn, NH, 2022).

______, 2022d, *HyperCosmos Fractionation and Fundamental Reference Frame Based Unification: Particle Inner Space Basis of Parton and Dual Resonance Models* (Pingree Hill Publishing, Auburn, NH, 2022).

______, 2022e, *A New UniDimension ProtoCosmos and SuperString F-Theory Relation to the HyperCosmos* (Pingree Hill Publishing, Auburn, NH, 2022).

______, 2022f, *The Cosmic Panorama: ProtoCosmos, HyperCosmos, Unified SuperStandard Theory (UST) Derivation* (Pingree Hill Publishing, Auburn, NH, 2022).

______, 2022g, *Ultimate Origin: ProtoCosmos and HyperCosmos* (Pingree Hill Publishing, Auburn, NH, 2022).

______, 2023a, *UltraUnification and the Generation of the Cosmos* (Pingree Hill Publishing, Auburn, NH, 2023).

INDEX

About The Author

Stephen Blaha is a well-known Physicist and Man of Letters with interests in Science, Society and civilization, the Arts, and Technology. He had an Alfred P. Sloan Foundation scholarship in college. He received his Ph.D. in Physics from Rockefeller University. He has served on the faculties of several major universities. He was also a Member of the Technical Staff at Bell Laboratories, a manager at the Boston Globe Newspaper, a Director at Wang Laboratories, and President of Blaha Software Inc. and of Janus Associates Inc. (NH).

Among other achievements he was a co-discoverer of the "r potential" for heavy quark binding developing the first (and still the only demonstrable) non-Aeolian gauge theory with an "r" potential; first suggested the existence of topological structures in superfluid He-3; first proposed Yang-Mills theories would appear in condensed matter phenomena with non-scalar order parameters; first developed a grammar-based formalism for quantum computers and applied it to elementary particle theories; first developed a new form of quantum field theory without divergences (thus solving a major 60 year old problem that enabled a unified theory of the Standard Model and Quantum Gravity without divergences to be developed); first developed a formulation of complex General Relativity based on analytic continuation from real space-time; first developed a generalized non-homogeneous Robertson-Walker metric that enabled a quantum theory of the Big Bang to be developed without singularities at t = 0; first generalized Cauchy's theorem and Gauss' theorem to complex, curved multi-dimensional spaces; received Honorable Mention in the Gravity Research Foundation Essay Competition in 1978; first developed a physically acceptable theory of faster-than-light particles; first derived a composition of extremums method in the Calculus of Variations; first quantitatively suggested that inflationary periods in the history of the universe were not needed; first proved Gödel's Theorem implies Nature must be quantum; provided a new alternative to the Higgs Mechanism, and Higgs particles, to generate masses; first showed how to resolve logical paradoxes including Gödel's Undecidability Theorem by developing Operator Logic and Quantum Operator Logic; first developed a quantitative harmonic oscillator-like model of the life cycle, and interactions, of civilizations; first showed how equations describing superorganisms also apply to civilizations. A recent book shows his theory applies successfully to the past 14 years of history and to *new* archaeological data on Andean and Mayan civilizations as well as Early Anatolian and Egyptian civilizations.

He first developed an axiomatic derivation of the form of The Standard Model from geometry – space-time properties – The Unified SuperStandard Model. It unifies all the known forces of Nature. It also has a Dark Matter sector that includes a Dark ElectroWeak sector with Dark doublets and Dark gauge interactions. It uses quantum coordinates to remove infinities that crop up in most

interacting quantum field theories and additionally to remove the infinities that appear in the Big Bang and generate inflationary growth of the universe. It shows gravity has a MOND-like form without sacrificing Newton's Laws. It relates the interactions of the MOND-like sector of gravity with the r-potential of Quark Confinement. The axioms of the theory lead to the question of their origin. We suggest in the preceding edition of this book it can be attributed to an entity with God-like properties. We explore these properties in "God Theory" and show they predict that the Cosmos exists forever although individual universes (or incarnations of our universe) "come and go." Several other important results emerge from God Theory such a functionally triune God. The Unified SuperStandard Theory has many other important parts described in the Current Edition of *The Unified SuperStandard Theory* and expanded in subsequent volumes.

Blaha has had a major impact on a succession of elementary particle theories: his Ph.D. thesis (1970), and papers, showed that quantum field theory calculations to all orders in ladder approximations could not give scaling deep inelastic electron-nucleon scattering. He later showed the eigenvalue equation for the fine structure constant α in Johnson-Baker-Willey QED had a zero at $\alpha = 1$ not 1/137 by solving the Schwinger-Dyson equations to all orders in an approximation that agreed with exact results to 4th order in α thus ending interest in this theory. In 1979 at Prof. Ken Johnson's (MIT) suggestion he calculated the proton-neutron mass difference in the MIT bag model and found the result had the wrong sign reducing interest in the bag model. These results all appear in Physical Review papers. In the 2000's he repeatedly pointed out the shortcomings of SuperString theory and showed that The Standard Model's form could be derived from space-time geometry by an extension of Lorentz transformations to faster than light transformations. This deeper space-time basis greatly increases the possibility that it is part of THE fundamental theory. Recently, Blaha showed that the Weak interactions differed significantly from the Strong, electromagnetic and gravitation interactions in important respects while these interactions had similar features, and suggested that ElectroWeak theory, which is essentially a glued union of the Weak interactions and Electromagnetism, possibly modulo unknown Higgs particle features, be replaced by a unified theory of the other interactions combined with a stand-alone Weak interaction theory. Blaha also showed that, if Charmonium calculations are taken seriously, the Strong interaction coupling constant is only a factor of five larger than the electromagnetic coupling constant, and thus Strong interaction perturbation theory would make sense and yield physically meaningful results.

In graduate school (1965-71) he wrote substantial papers in elementary particles and group theory: The Inelastic E- P Structure Functions in a Gluon Model. Phys. Lett. B40:501-502,1972; Deep-Inelastic E-P Structure Functions In A Ladder Model With Spin 1/2 Nucleons, Phys.Rev. D3:510-523,1971; Continuum Contributions To The Pion Radius, Phys. Rev. 178:2167-2169,1969; Character Analysis of U(N) and SU(N), J. Math. Phys. <u>10</u>, 2156 (1969); and The Calculation of the Irreducible Characters of the Symmetric Group in Terms of the

Compound Characters, (Published as Blaha's Lemma in D. E. Knuth's book: *The Art of Computer Programming Vols. 1 – 4*).

In the early 1980's Blaha was also a pioneer in the development of UNIX for financial, scientific and Internet applications: benchmarked UNIX versions showing that block size was critical for UNIX performance, developing financial modeling software, starting database benchmarking comparison studies, developing Internet-like UNIX networking (1982) and developing a hybrid shell programming technique (1982) that was a precursor to the PERL programming language. He was also the manager of the AT&T ten-year future products development database. His work helped lead to commercial UNIX on computers such as Sun Micros, IBM AIX minis, and Apple computers.

In the 1980's he pioneered the development of PC Desktop Publishing on laser printers and was nominated for three "Awards for Technical Excellence" in 1987 by PC Magazine for PC software products that he designed and developed.

Recently he has developed a theory of Megaverses – actual universes of which our universe is one – with quantum particle-like properties based on the Wheeler-DeWitt equation of Quantum Gravity. He has developed a theory of a baryonic force, which had been conjectured many years ago, and estimated the strength of the force based on discrepancies in measurements of the gravitational constant G. This force, operative in D-dimensional space, can be used to escape from our universe in "uniships" which are the equivalent of the faster-than-light starships proposed in the author's earlier books. Thus travel to other universes, as well as to other stars is possible.

Blaha also considered the complexified Wheeler-DeWitt equation and showed that its limitation to real-valued coordinates and metrics generated a Cosmological Constant in the Einstein equations.

The author has also recently written a series of books on the serious problems of the United States and their solution as well as a book on the decline of Mankind that will follow from current social and genetic trends in Mankind.

In the past twenty years Dr. Blaha has written over 80 books on a wide range of topics. Some recent major works are: *From Asynchronous Logic to The Standard Model to Superflight to the Stars, All the Universe!, SuperCivilizations: Civilizations as Superorganisms, America's Future: an Islamic Surge, ISIS, al Qaeda, World Epidemics, Ukraine, Russia-China Pact, US Leadership Crisis, The Rises and Falls of Man – Destiny – 3000 AD: New Support for a Superorganism MACRO-THEORY of CIVILIZATIONS From CURRENT WORLD TRENDS and NEW Peruvian, Pre-Mayan, Mayan, Anatolian, and Early Egyptian Data, with a Projection to 3000 AD*, and *Mankind in Decline: Genetic Disasters, Human-Animal Hybrids, Overpopulation, Pollution, Global Warming, Food and Water Shortages, Desertification, Poverty, Rising Violence, Genocide, Epidemics, Wars, Leadership Failure.*

He has taught approximately 4,000 students in undergraduate, graduate, and postgraduate corporate education courses primarily in major universities, and large companies and government agencies.

He developed a quantum theory, The Unified SuperStandard Theory (UST), which describes elementary particles in detail without the difficulties of conventional quantum field theory. He found that the internal symmetries of this theory could be exactly derived from an octonion theory called QUeST. He further found that another octonion theory (UTMOST) describes the Megaverse. It can hold QUeST universes such as our own universe. It has an internal symmetry structure which is a superset of the QUeST internal symmetries.

Recently he developed Octonion Cosmology. He replaced it with HyperCosmos theory, which has significantly better features. He developed a fractionalization process for dimensions, particles and symmetry groups. He also described transformation that reduced particles and dimensions to a far more compact form. He also developed a precursor theory ProtoCosmos that leads to the HyperCosmos.

The author showed that space-time and Internal Symmetries can be unified in any of the ten HyperCosmos spaces in their associated HyperUnification spaces. The combined set of HyperUnification spaces enable all HyperCosmos dimensions to be obtained by a General Relativistic transformation from one primordial dimension in the 42 space-time dimension unified HyperUnification space.